GOETHE ON NATURE

&

ON SCIENCE

GOETHE ON NATURE
&
ON SCIENCE

BY

SIR CHARLES SHERRINGTON, O.M.

THE PHILIP MAURICE DENEKE LECTURE
DELIVERED AT
LADY MARGARET HALL, OXFORD
ON THE 4TH MARCH 1942

CAMBRIDGE
AT THE UNIVERSITY PRESS
1949

CAMBRIDGE
UNIVERSITY PRESS

University Printing House, Cambridge CB2 8BS, United Kingdom

Published in the United States of America by Cambridge University Press, New York

Cambridge University Press is part of the University of Cambridge.

It furthers the University's mission by disseminating knowledge in the pursuit of education, learning and research at the highest international levels of excellence.

www.cambridge.org
Information on this title: www.cambridge.org/9781107652675

© Cambridge University Press 1949

First edition 1942
Second edition 1949
First published 1949
First paperback edition 2014

A catalogue record for this publication is available from the British Library

ISBN 978-1-107-65267-5 Paperback

GOETHE ON NATURE

&

ON SCIENCE

TWO facts we may, at outset, recall about Goethe. Poet though he was, he was yet life-long an ardent student of the sciences of Nature. And this other, that with him—not merely as usage of the German language—Nature was usually Nature with a capital N. The thoughts of few men can be more liberally on record than are Goethe's, biographically and autobiographically, in his formal works and in his correspondence. We may look therefore to exceptional opportunity for knowing what this Nature with a capital N stood for in his mind.

He would have wished us to know. He was disappointed that his contemporaries did not pay more attention to his thoughts on Science and on Nature. He remarked more than once: 'I do not attach importance to my work as a poet, but I do claim to be alone in my time in apprehending the truth about colour.'[1] Again, in the pillage of Weimar, his main anxiety was for his scientific work in manuscript.[2]

We know something of his reaction to Nature in his childhood. A boy, during the family's theological dis-

[1] *Gespräche m. Eckermann* (Oxenford trans.; Dent and Co.), p. 302.
[2] Duntzer, *Life*, p. 745.

cussions, he would sit silent, but on occasion in his bedroom afterwards would contrive an altar from a music-stand, decked with minerals and flowers, and crowned by a flame lit by a burning glass from the rays of the freshly risen sun. Clearly, a child's act of worship. Paganism we may think, but the boy came of a zealous Lutheran family.

In manhood, scientific studies always interested him. Let me use the term science here in brief to mean the sciences of Nature. Goethe's science, though not profound, was broad and reached physics, geology and biology.

Goethe was an assiduous investigator. In Rudolph Magnus's attractive volume *Goethe als Naturforscher* ★ all that side of him is charmingly set forth. He gave hours to observation and to experiments of a simple kind, covering all types of life as well as those of the 'higher' plants and animals. In physics, his best-known work—he called it 'optical'—was on colour. A feature of it was disagreement with that fundamental observation, already established a hundred years, known as the decomposition of white light into coloured lights on passing through a glass prism. He rejected this. He took exception even to the expression coloured 'lights' because he said, there is only one light.[1] He demurred against the new 'abstract' word 'refractivity'.[2] He asked, what does it mean?

In this same matter of coloured and white light, he found himself likewise unable to confirm that white could

[1] *Gesammte Werke*, Goethe, vol. XXVIII, p. 296. Cotta, Stuttgart, 1858.   [2] Vol. XXVIII, p. 297.

★ Here and elsewhere an asterisk indicates that there is a note on pp. 51 ff.

be arrived at by combining colours. 'Dass alle Farben, zusammengemischt, Weiss machen, ist eine Absurdität (1808).¹' [That all colours mixt together make white is an absurdity.]

In his *Annalen* for 1790 Goethe noted that, while thinking over the principles of Painting, he found 'to his astonishment' that Newton's work on colour was demonstrably wrong. On turning to the *Life*² we find a more circumstantial account. Büttner the botanist had on Goethe's request lent him some prisms in order to perform the experiments of Newton. But time slipped and Goethe never unpacked them—the prisms lay still in their box when the anatomist Loder wrote that the owner of the prisms was getting unhappy at receiving neither response nor prisms from Goethe. The prisms were at last unpacked, and before sending them off to their owner it occurred to Goethe to take one look through a prism. To his amazement the white wall at which he gazed through the prism remained white. Colour showed only where something dark edged the white. Colour showed brightest of all on the window frames. Goethe immediately concluded that he had thus discovered the Newtonian account of light to be an error.

In the following year³ he published what has been styled satirically '*Goethe* v. *the prism*'. For himself he spoke of it as the 'Newton controversy'. In view of the known character of Newton it seems unlikely that had Newton been alive—he had been dead 70 years—he

¹ *Entw. ed. Farbenlehre*, par. 558, 28, p. 146.
² Duntzer, *Life*, bk. VI, cap. 2, p. 441.
³ *Beit. z. Optik.* (Weimar, 1791 and 1792).

would have returned to a discussion he looked upon as settled. But despite having no living Newton to tilt against Goethe entered upon the wordy campaign which lasted the rest of his life. In it Goethe would, we must remember, appear to many not as the aggressor. To many he would stand as the protagonist of ancient orthodoxy, following as he did in the broad wake of Aristotle and Theophrastus. He translated Theophrastus (Aristotle) on Colour (1801). In 1824 he was imparting to Eckermann the Aristotelian teaching as still current! It was Newton with his prisms (1704,[1] in fact even earlier, 1671) whom Goethe could treat as the rebel. Moreover, Goethe held certain views as to what a scientific experiment should be, and the prism experiment did not conform with them. The prism introduced 'hundertlei'[2] complications, and dragged in mathematics unwantedly. They had been introduced by a mathematician (Newton—Goethe seems habitually to have thought of Newton less as a natural philosopher than as a mathematician), though they were not part of the subject. The prism was an extraneous accessory. With Goethe the prism stood for 'mathematics'. Goethe argued too that the prism implied a naïve attempt to analyse not colour but light itself. 'Light', he said, 'is an elemental entity, and inscrutable attribute of creation, an "Einziges",[3] which has to be taken for granted.' To try to analyse light was a shallow blunder. And the manner of the attempt! Through a tiny hole to admit a poverty-

[1] *Opticks or a Treatise of the Reflexions, Refractions, Inflexions and Colour of Light* (London, 1704).

[2] *Ges. Werke*, vol. XXVIII, p. 296, par. 25.

[3] Ibid.

8

stricken thread of light into a darkened room, when by going into the open day any amount of it could be had— no wonder the students laughed[1] and ran off! And the supposition to which it led! A physicist's mathematics! There Goethe may have thought he could hear the chuckle of Mephistopheles himself. Streams of travelling particles or wavelets pouring from the sun! Why, we had but to step into the free daylight to see that it was not so. We encounter here in Goethe what seems an almost wilful inability to enter into the physicist's point of view. As was his wont, when thoroughly enjoying himself he broke into verse about this:

> Möget ihr das Licht zerstückeln,
> Farb' um Farbe draus entwickeln,
> Oder andre Schwänke führen,
> Kügelchen polarisiren.
> Dass der Hörer ganz erschrocken
> Fühlet Sinn und Sinne stocken.
> Nein! es soll euch nicht gelingen.[2]

These lyrical outbursts illustrate an artistic principle which Goethe pressed. He said 'The world is so great and rich, and life so full of variety, that you can never want occasions for poems. But they must all be *occasioned*; that is to say, reality must give both impulse and material! All of my poems are occasioned poems, suggested by real life, and having therein a firm foundation. I attach no value to poems snatched out of the air.'[3]

[1] *Ges. Werke*, vol. XXX, p. 12: *Nachtr. zur Farbenl.*
[2] *Gesammte Gedichte*, vol. II, p. 147: *Gott u. Welt.*
[3] *Gespräche m. Eckermann* (1823), p. 8.

The *Annalen* relate how in 1817 he was present at a demonstration of the corpuscular theory of light given at Court. He writes of it that 'a university professor' 'with incredible composure and assurance flourished the most impudent trickery before high (i.e. in the Court of Weimar) and intelligent people. After gazing and gazing, after blinking and blinking (with aching eyes), you were quite at a loss to know either what you had seen or what you were intended to see. At the first preparations I got up and went off, and on my return heard without surprise the course of his demonstration, as I had foreseen it. I was also taught on this occasion, by the illustration of billiard balls, how the round molecules of light, if they strike the glass with the poles, penetrate quite through, whereas if they meet it with the equator they are sent back with protest.'[1] Protest! How characteristically Goethean and medieval!

Goethe complained, when trying to repeat some of the Newtonian experiments with a view to publication of his own *Farbenlehre*, 'that by joining together several instruments' (optical) Newton has 'perpetrated an experimental incoherence'.[2] But, enjoying recently the Tercentenary Lecture (1942) on Newton by our distinguished physicist and authority on Newton, Professor Andrade, I note that, after critically evaluating Newton's optical researches, he adds: 'Newton's work on light establishes him as a supreme experimenter'.[3]

[1] Translated by Charles Nisbet; George Black and Sons, London 1892, p. 437.

[2] *Annalen*, p. 363 (for year 1807).

[3] *Proceedings Physical Society* (1943), vol. LV, p. 129, London.

Condemning though he did Newton's experiment 'in unmeasured terms'[1] Goethe yet, through all the years, was never at the pains to repeat it.[2] He said that prisms are 'always somewhat clouded', thus suggesting that the colour effect of a prism is due, as he contended all physical colour is, to cloudiness of a semi-opaque medium. He declared that to suppose all light is not one and the same and to suppose white light contains coloured light is to suppose absurdities. But criticising the point Helmholtz expressly states he does not understand in what way they appear absurd, and that Goethe does not declare in what way they do so.

And what did Goethe, rejecting Newton's fact of refraction and colour, put in its place? The following is his substitute. The phenomena of Nature, he tells us, are of two grades. The majority do not lend themselves well to analysis because in them the fundamental is obscured by the accessory. There are, however, certain natural phenomena which do lie open to human inquiry in their naked simplicity. This latter class are *Urphänomenen* or ground-phenomena.[3] The Urphänomen is fundamental in significance.

Magnetism and the magnet exemplify it. It is something '*das man nur aussprechen darf um es erklärt zu haben*', i.e. is self-explanatory. We comprehend it instinctively. Science cannot, and never will, resolve further an Urphänomen. But by it a foundation is given on which to build. It allows insight into Nature. Thus, magnetic attraction and

[1] Helmholtz's *Physiologische Optik* (1856–66), p. 267.
[2] Ibid. p. 268.
[3] *Zur Farbenlehre*, vol. II, pars. 174–7.

repulsion 'zusammen deuten auf eine Scheidung, auf eine Entzweien das beim Magnet sein Entgegengesetzten, seine Totalität, sein Ganzes er wiedersucht'. Goethe declares that there 'is in Nature, both animate and inanimate, a something which manifests itself as contradiction'.[1]

This smacks of Hegel,[2] but Goethe would hardly have admitted that. He was friendly with Hegel the man, but avoided Hegel the sophist.

In Mineralogy and Geology the Urphänomen is granite, because granite lies at the base of the Earth's crust; the very heart of the mountains consists of it, 'my spirit's wings can go no further'.[3]

Goethe continues:

Treue Beobachter der Natur, wenn sie auch sonst noch so verschieden denken, werden doch darin mit einander übereinkommen, das alles, was erscheinen, was uns als ein Phänomen begegnen solle, müsse entweder eine ursprünglicher Entzweiung, die einer Vereinigung fähig ist, oder ursprünglich Einheit, die zur Entzweiung gelangen könne, andeuten, und sich auf eine solche Weise darstellen. Das Geeinte zu entzweien, das Entzweite zu einigen, ist das Leben der Natur: dies ist die ewige Systole und Diastole, die ewige Synkrisis und Diakrisis, das Ein- und-Ausathmung der Welt, in der wir leben, weben und sind.[4]★

[1] *Ges. Werke*, vol. xx, p. 156: *Wahrh. u. Dicht.*

[2] *Gespräche m. Eckermann*, p. 244.

[3] *Ueber den Granit* (1782). Cf. *Gespräche m. Eckermann* (18 May 1824), p. 65. Everyman's Library.

[4] *Ges. Werke*, vol. xxviii, Abth. 4, par. 739, p. 187: *Zur Farbenlehre.*

A rhapsody!

The conception 'Urphänomen' as applicable truly to Nature, animate or inanimate, has failed, in spite of Goethe's advocacy, to establish itself in Science. More true to scientific thought has been the little quatrain from Tennyson:

> Flower in the crannied wall
> I pluck you out of the crannies,
> And if I could understand you all in all
> I should know what God and Man is.

That at least puts to us tellingly how Nature, despite all its complexity and its more than million-fold detail, is one unified system; though perhaps we may owe this impression of the unity of Nature to the circumstance that as observers we each, taken singly, observe as an integrated unit.

Goethe was one of those who, relying on introspection, believe that green as perceived contains 'perceived yellow and blue', an assertion difficult to challenge though difficult to accept. Arthur Schopenhauer supported warmly part of Goethe's views on colour; Goethe's *Annalen* mentions that, meeting in 1816, they agreed partly to disagree; but Schopenhauer, on publishing his book, speaks of Goethe taking him through the experiments in 1813. Goethe asserts that Hegel became an adherent to his view —but his adherence was temporary.[1] In later years, Tyndall, Arthur Schuster and W. Ostwald* took up the cudgels, and as physicists. Goethe's own description of the 'turbidity tints' suggests that either they were for him stronger than

[1] *Annalen*, p. 438.

13

for most observers or that he overstated them. Dealing with Goethe's description as given in the *Farbenlehre* Helmholtz in the *Handbuch*[1] (1860), after furnishing a précis of Goethe's account, writes 'This description of the matter, if intended for physical, has no sense,...'. 'These Goethean descriptions are not to be understood as physical, but only as figurative dramatizations [or "interpretations"] of the process (*als bildliche Versinnlichungen des Vorgangs*).'[2] In other words, Helmholtz declined to accept the Goethe lucubrations as lying within the province of science at all. I can only say, in my later generation, I agree entirely with Helmholtz on the point.

Goethe never reconciled himself* to the Newtonian notion of light, though it became the stand-by on the subject for the scientific world. Perhaps as a corollary to that dislike, never does he, although writing and thinking about colour practically up to his death (1833), mention Thomas Young's theory (1801) which simplified the mathematical treatment of colour sensations by postulating three primary sensations due to sufficiently separated wave-lengths, a supposition later adopted by Helmholtz, and used as a working hypothesis for a century and a half. Newton had separated and dealt with a sample of the electro-magnetic radiation later assumed by physicists to stream from universe to universe; Young had taken advantage of that to examine physically Nature's great gift to man of colour. Goethe would listen to none of these 'physical' fancies. Refraction and polarisation? he said, 'beide sind hohle Worte die Denkenden gar

[1] *Physiologische Optik*, 1856–66, and again 2nd ed.
[2] *Physiologische Optik*, 1856–66, and in successive editions.

nichts sagen, und die doch so oft von wissenschaftlichen Männern wiederholt werden!'[1]

In following Goethe's 'science' we are helped by his having laid down principles which in his view should govern scientific observation. One of them is that the conditions for observation be kept as simple as possible, and for that reason should eschew apparatus. Prominent in his objection to the prism experiment was that the prism introduced heaven-knows-what complications. Essential for scientific observation was *Anschaulichkeit* 'obviousness' or 'naked clarity'. This clarity could dispense with mathematics. Goethe was not himself equipped in mathematics,[2] and he regarded the role of mathematics in science with distrust. Mathematics led to the introduction of propositions which were not truly contained in the original proposition.[3] They had brought calamity to optics. He did not see that a use of apparatus is to simplify conditions. Nor again, that mathematics can be a main means toward obtaining *Anschaulichkeit*.

Goethe was, it is well to remember, of that type which the psychologist classes as 'visual'. That is, his memories, his fancies, his dreams, used visual imagery. To induce sleep he would imagine a seed gradually growing into a plant. His predilection for the eye is expressed by his remark 'gegen das Auge ist das Ohr ein stummer Sinn'.

The late Rudolf Magnus, to whose book I am deeply indebted in this short essay, gave much and sympathetic attention to Goethe's work on colour, and he argued that

---

[1] *Ges. Werke*, vol. XXX, p. 394 and vol. XXVIII, p. 297.

[2] Ibid. vol. XXVIII, p. 183.

[3] Ibid. vol. XXX, p. 385.

there were times when Goethe did not succeed in bearing in mind that the physical, although it correlates with the psychical, need not resemble it.

In his study of colour Goethe took as his starting-point what he regarded as its Urphänomen. That was this. If a little spirits-of-wine containing a trace of soap be added to a glass of water, the clear water becomes clouded. Held up before a light, that is, seen by light coming through it, the water appeared yellowish. On the other hand, lit by the same light from the front, and a black screen behind to cut off transmitted light, the clouded water looks bluish. Goethe nowhere offers an explanation of these colour-effects of turbidity; he treats them simply as empirical facts. The blue of the sky is now ascribed to scattering of light by small particles, perhaps even air molecules. The shorter (blue) wave-lengths are much more affected than are the longer. That is, when the light by which we see it comes through it, the water, with its faint uncoloured turbidity, appears yellowish. When, using the same white light, the light is reflected from it, there is a bluish colour. Probably most of us have noticed a similar effect when a little milk is added to water; a richer emulsion, for instance undiluted milk, is fully opaque to light. We have then for transmitted light complete extinction, darkness. And to reflected light a surface which, like the familiar appearance of milk, is white because fully reflecting. This final degree of clouding, complete opacity, was for Goethe 'vollendete Trübe'. 'Turbidity...is the initial rudiment whence is developed the whole science of chromatics.'[1] This ground-pheno-

[1] *Annalen* (1806), p. 348.

16

menon, like that of the magnet, had the property of exhibiting diametrical opposites, white and black, brightness and darkness, yellow and blue; between these two polar opposites stretched a series of the colours.

Colour is a problem involving a considerable range of scientific technique. Over and above those extensions of it, such as aesthetics, which are purely psychological, it involves on the one hand light, the physical process as first set forth by Newton, on the other hand, visual sensation as studied by physiology and psychology. It is a problem, therefore, which strides back and forth across a common zone where physical and psychical overlap. The technical terms lie somewhat open to confusion.

The turbidity colours were, according to Goethe, simplicity itself. They need no apparatus. They invite no sophistication at the hands of mathematics. They are immediately understandable by all. They are *Anschaulichkeit* pure. With this as his Urphänomen Goethe developed his 'Doctrine of Colour', 'die Farbenlehre'.[1] Its teaching was: We see through media. Colour is ultimately an affair of the cloudinesses of media. Goethe had 'opalglasses' prepared (still extant at Weimar), which gave by reflected light a bluish tint, by transmitted light a yellow or orange. He drew confirmation and illustration of the theory from an incident he heard. An old oil-painting, of a man dressed in black, was being cleaned with a wet sponge. At once the black coat looked blue, but when dry was black again. The surface sheet of varnish, clouded by moisture and seen by reflected light against the black ground, appeared blue. The harvest moon through

[1] *Ges. Werke*, vol. XXIX, p. 322.

autumnal mist flames like a blood orange. Then there is the roseate glow of clouds at sunrise and sunset, and above that glow the bar of heavy purple. Then again the blueness of the hills of the far landscape. Again, the immaculate blue of the summer zenith, which is the gauze of scattered light from the unclouded air above us seen against the intense black of cosmic space.[1] Goethe is so happy about all this that he breaks into verse.

> Wenn der Blick an heitern Tagen
> Sich zur Himmel's Blaue lenkt,
> Beim Siroc der Sonnenwagen
> Purpurrot sich nieder senkt;
> Da gebt der Natur die Ehre
> Froh, an Aug' und Herz gesund,
> Und erkennt der Farbenlehre
> Allgemeinen, ew'gen Grund.[2]

He went on to observations in which he could be freer still from preoccupation with the physical notion of light. We all have experience of, after looking at a strong colour, seeing for a moment the contrast-colour of the original. On a bright morning a red rose will repeat itself in green on the gravel path. Goethe studied, as had Leonardo and others, the occurrence of these after-images. He arranged the colours in a ring of six—three which he considered simple colours and three intervening which he considered were mixed ones. In this ring the colour of the after-image always stood opposite to the colour which induced it. This result he likened to the Urphänomen of the

[1] *Ges. Werke*, vol. xxx, p. 20: *Falsche Ableitung des Himmel-blauen.* Cf. Brewster, Stokes, fluorescence.

[2] *Ges. Gedichte*, vol. II, p. 147: *Gott u. Welt*; *Gesetz d. Trübe.*

magnet. It showed Nature striving to satisfy the longing for totality when forced into polar opposites. This phenomenon is known as 'successive contrast'. There is also 'simultaneous contrast'. A sample is this. Two sheets of paper, one red, one green, are placed out on the table in a good light; a bit of pure grey paper laid on the red sheet looks green, on the green sheet looks red. Red and green are polar opposites; the calling forth of one of them by itself is an 'Entzweiung' of the 'Ganzes' in one part of the field, and 'Totalität' makes the other appear even where not called up in the field. Goethe remarked of this phenomenon that it arises unconsciously. Later work[1] here fully confirmed Goethe—the contrast we experience arises from quite unconscious mental factors. It dogs the painter all his time—and Goethe was as skilled as a painter in observing it. He speaks of the waves of a green sea adding purple to their moving shadows. He enjoyed talking of colour with painters though it is true he had not contact with any of the greater colourists.

This predominance of the visual in him was evident in his scientific studies, dealing with the twin kingdoms of living nature, plant and animal. It was in the shapes assumed by life that Goethe found his fullest scientific interest—an interest which never staled in him. Aristotle is known as the founder of the study. But Goethe gave it the name by which it is now universally known, morphology—the study of shape, meaning by that living shape. The subject attracted him early. When a Law student at Strasburg he preferred the company of the

[1] Cf. Sherrington, C. S., *J. Physiol.* (1896), vol. xx; (1897), vol. xxi, pp. 33–5.

students who were doing anatomy, because they talked 'shop' even at meals. He frequented the dissections. The human body is of course a treasure-house of organized form. Later, at Weimar, the shapes of trees and leaves fascinated him. Going by way of the Brenner and of Garda, he found Italy an enchanted land. As he descended the southern alpine slope the wealth of flower and leaf and stem was something he had never witnessed or conceived. At Padua he entered the Botanical Garden, where the palms were in blossom. Their fans illustrated a scientific thought he had entertained, their shapes presenting a series ranging from extreme to extreme with all grades of transition. This suggested to him that leaf-form was in a state of flux.

Superintending State forestry and agriculture at Weimar as he did, Goethe was conversant with the botany of his time. He would take with him the young Dietrich on excursions for collecting plants. The Linnean system of classification was everywhere in use, and young Dietrich knew it well. Goethe had difficulty in memorizing the names. He blamed this to the system. His poem to Christiane Vulpius complains:

Dich verwirret, Geliebte, die tausendfältige Mischung
  Dieses Blumengewühls über dem Garten umher;
Viele Namen hörest du an, und immer verdränget
  Mit barbarischem Klang einer den andern im Ohr.[1]

Botany was an old study, and in the seventeenth century had at last surmised the meaning of the flower of the plant. There had arisen a desire to catalogue, and arrange in

[1] *Ges. Gedichte*, vol. II, p. 140: *Die Metam. d. Pflanzen.*

order, all the kinds of plants that were, the assumption of the time being that the kinds around us were still as at the Creation of the world. Goethe, however, had noticed for himself such differences between specimens of the same species that he thought specific form might be in a state of flux. The continual change, which he suspected was going on in plant-form, Goethe attributed to the influence of external conditions—soil, light, warmth, moisture, etc.

The naturalist Lamarck was putting forth somewhat similar views in Paris. Both Goethe and he dissented from the Linnean 'frozen' view of species, though it was orthodox. Lamarck is as prosy a writer as one can find; Goethe, on the other hand, sang his thesis aloud.

> Und umzuschaffen das Geschaffne
> Damit sich's nicht zum Starren waffne,
> Wirkt ewiges, lebendiges Tun.
> Und was nicht war, nun will es werden,
> Zu reinen Sonnen, farbigen Erden,
> In keinem Falle darf es ruh'n.
>
> Es soll sich regen, schaffen, handeln,
> Erst sich gestalten, dann verwandeln;
> Nur scheinbar steht's Momente still.
> Das Ewige regt sich fort in allen!
> Denn alles muss in Nichts zerfallen
> Wenn es im Sein beharren will.[1]

In reading Goethe's science we are never left long without a reminder of his tendency to personify Nature.

With this state of flux, which Nature's living work exhibited, the collateral notion grew upon him that, at

[1] *Ges. Gedichte*, vol. II, p. 138: *Gott u. Welt*; *Eins u. Alles*.

back of it all, were ideals—for instance that, while creating leaves, Nature kept in mind an 'ideal' leaf. Concrete leaves, in all their vast variety, were variants of an ideal leaf. His fancy pictured an 'ideal' plant, and Nature calling forth from the stem of it a manifold of side-growths, of leaves, petals, sepals, stamens, each and all of them just modifications of the ideal leaf. The very wrappings of the seed, the shell of the nut, the flesh of the apple, were all modifications of the leaf. As efficient causes actually producing these he supposed phases of growth due to restraint and freedom of the movement of the sap.

Such a surmise remains a surmise and nothing more, until fact verifies it. For this last in this case a series of observations would be needed, preferably we may think tracing stages in the development of the plant, the examinations of lateral outgrowths from the initial stem, and whether these have unequivocal properties of, or serve unequivocally as, leaves. Now the study of the development of a plant, or animal, is at root an affair of following its cellular development. The cell-theory had not arrived in Goethe's time. His suggestion that all the appendages of the plant stem are essentially leaves remained, therefore, perforce an uncritical assertion. When later the progress of botany in due course obtained facts competent for the question, the theory was found not to be borne out. It fell, therefore, into the doleful category of unlucky guesses.

Goethe regarded this conception as a botanical truth of first importance. In his enthusiasm he took occasion, almost at the beginning of their acquaintance—which ripened into close friendship—to dilate on this theme to Schiller. Schiller heard him out, and then, to Goethe's surprise, said: 'Das ist keine Erfahrung; das ist eine Idee.' —'That is not a fact: it is an idea.'[1] Time has confirmed Schiller's remark. Goethe's view has proved untenable as fact.

This whole incident is discussed fully and admirably by Magnus.[2] It makes an odd picture. Goethe, great master of language, consulting Schiller, another such master, as to whether a plausible guess is the same thing as an observation of fact. Goethe elsewhere remarks that Schiller was more a man of the world than he. In some things, perhaps, but Goethe was the more *rusé*. Goethe had been flying a kite, but Schiller with characteristic perspicuity and frankness declined to see it. Years later, in 1807, Goethe repeated the same sort of 'exploitation';[3] he refers to his conjecture about the skull being a piece of backbone as though it were a settled and accepted fact, whereas it was only a surmise and was unconfirmed by later knowledge.

Goethe studied animal form also. An anatomical finding which he came across was that the incisor part of the upper jaw in man is at first separate from the rest of that bone, as it is in animals. It was a detail, and Goethe was not in the least the type of naturalist who loves detail for detail's sake. But on his coming across this old vestigial partition of the human face-bone he writes to Herder: 'I have found—not gold or silver—but something which gives me unspeakable delight. I was comparing with Loden [anatomist] the human and animal skull, and came on the

---

[1] *Annalen* (1794), p. 210; cf. ibid. (1807), p. 361.
[2] *Goethe als Naturforscher*, p. 88.   [3] *Annalen*, p. 361.

clue.' His delight lay in the detail as a clue, one clue the more, bringing man and animal together. The tendency and wish of the time—under theological bias—was to put man and animal further apart. As for the observation as a discovery, it was in fact already known. It had been published a few years before by Vicq-d'Azyr in Paris, at the Académie des Sciences (1779).

Again in the animal world it is individual form which most attracts him—he classified even the shapes of clouds. As among plants the flowering plants, so among animals it was the crowning group—the back-boned—which specially engaged him. The spine is a row of bones, the vertebrae; jointed to its front end of the row is the skull. Goethe surmised that the skull itself is vertebrae continuing those of the back-bone. Goethe's musings conceived creative Nature while creating the back-boned animal, keeping in mind an ideal vertebra. He exclaims: 'How far from the tortoise to the elephant, and yet the gap is bridged entirely by intermediate forms! Because the whole series belongs to one ideal type.' The Nature-goddess, in shaping each great animal type, works to the pattern of a 'vorschwebende Idee'. This is in the fashion of the so-called *Nature-philosophen* of the period. That the skull is a set of vertebrae, an anatomist, Oken, put forward independently of Goethe, and it had a following among anatomists for a time. But with the progress of anatomy it became discredited, and in the '60's of last century Thomas Huxley rejected it finally.

These ideal patterns which the creative principle set before itself were, so to say, Platonic ideas in the mind of the creative spirit. Goethe drew joy from the contemplation of the boundless productivity of this Being. Yet, deifying her though he did, he traced limits to the powers she possessed. He took credit to himself for the discovery of a 'law' which he called that of the 'correlation of parts'. It decreed that nothing could be added new to an animal-shape except at the cost of taking something away. Thus, the long body of the snake is obtained by depriving the creature of limbs. The relatively large limbs of the frog are got at the expense of shortening the body. Even the ateliers of Olympus are therefore under the rule of necessity. The deciphering of this law of correlation so pleased him that he broke into song about it.

Siehst du also dem einen Geschöpf besonderen Vorzug
Irgend gegönnt, so frage nur gleich, wo leidet es etwa
Mangel anderswo, und suche mit forschendem Geiste,
Finden wirst du sogleich zu aller Bildung den Schlüssel.
Denn so hat kein Tier, dem sämmtlich Zähne den obern
Kiefer umzäumen, ein Horn auf seiner Stirne getragen,
Und daher ist den Löwen gehört der Ewigen Mutter
Ganz unmöglich zu bilden, und böte sie alle Gewalt auf,
Denn sie hat nicht Masse genug, die Reihen der Zähne
Völlig zu pflanzen und auch Geweih und Hörner zu
    treiben.[1]

These verses are scarcely poetry. Their theme we may think does not fully admit of poetry. That Goethe himself should so misjudge the theme is poignant evidence of how greatly in earnest he was about it. The so-called 'law' is a mistake, and is no part of science to-day.

This elderly controversy over colour and how colour is produced, is out of date; it was settled a hundred years ago.

[1] *Ges. Gedichte*, vol. II, p. 143: *Die Metamorphose der Thiere.*

Why then revert to it now? Helmholtz, dismissing Goethe's *Farbenlehre* as not coming within the category of Science at all, goes on to say that in its time the great attention paid to it in Germany depended in part on the circumstance that the general public, unversed in strict scientific inquiries, inclined to picturesque presentation of the subject rather than to mathematico-physical abstractions.[1] The same situation obtains to-day, though not altogether in the same way. Goethe is by many[2] still supposed to be an authority in science, though not, perhaps, by those who have actually read his 'science'. Briefly, what was his 'science'? That light is *not* resoluble into coloured lights! An error reiterated to his life's end. That the plant is a collection of modified leaves; that the skull is an adapted piece of back-bone—two plausible though superficial conjectures, now, in the fuller light of the cell theory and embryology, set aside; the 'correlation of parts', a clumsy error he misthought a 'law'.

Occasion for controversy is therefore past and gone. Those who want can read the facts. That sanctions brevity here.

It was in Goethe's time that Johannes Muller put forward his 'law of specific energies'. 'It stated that any given nerve-fibre transmits a certain "quality" which is independent of the nature of the stimulus' (Granit). The fact that the rate of travel and other features of the action-potential (impulse) vary with coarseness of the fibre, offered a basis for differentiating the 'impulses' at least in the nerve-fibres of different thickness.

[1] *Physiologische Optik* (1860), p. 268.
[2] E.g. *A Study of Goethe*, by Professor Barker Fairley (1947).

But a nerve-path consists not of a single length of nerve-fibre but of a chain of such linked end-to-end—and at the junction of the links the incoming fibre breaks up into smaller fibres, so that the 'action-potential' loses there its earlier dimensions in space and time, and, following the 'principle of convergence', actually merges into others. If what is wanted for the due working of the nervous system is distinction as to which among its incoming fibres are active, that of course can be secured by means of a distinctive pattern attaching to the central activity excited. The central organ to which the incoming nerve-paths run is far better fitted for that than are the mere incoming paths themselves, and it is in accord with this very purpose that the central organ has been constructed and evolved. The central reaction supplies the 'local sign' even when the reaction is merely reflex. The spinal frog when touched brings its foot to the touched spot (local sign). A reflex nerve-path running from the skin-point to the muscles moving the foot accounts for that, as careful experiment has proved. How far this act has a psychical accompaniment none can say. It is plain, however, that the act as motor behaviour is a proto-type of what I might do myself. There is no impassable gap between that and my walking about perceiving the world I am in. 'Local sign' in a peripheral nerve-fibre is no longer demanded.

Since Magnus published, now 42 years ago, his review of Goethe's observations on colour, additions to know-ledge of the subject have been many. Professor Granit has introduced a micro-electrode to tap the electric disturbance in a retinal unit. At many points the response is restricted to a range of light almost monochromatic. Professor Granit finds also a brightness element partly dependent on colour and partly independent. The *dominator-modulator* theory, accounting for this, turns to the convergence of brightness (rod) and colour (cone) receptors upon the same common nerve-fibre, a fertile conception. Professor Hartridge, reviewing the effects of minimal retinal stimuli on himself, has reaffirmed that at threshold intensity these effects are colourless, etc., and that there seem to be several types of punctate colour-response in the normal human retina, the poly-chromatic theory. There are also the no less important recent explorations into the chemistry of visual purple (the Tansleys). None of all these observations are directed to the issues raised by Goethe against Newton, so we may pass them over here.

Zur Farbenlehre is a treatise of more than 450 octavo pages. It deals with aspects of its subject which are largely outside our particular interest here. 'On the allegorical, symbolic, and mythical use of colour', etc. It deals with some of these so trivially that the reader wonders what the intention could be. For instance, a section on the colours of animals, which might want a volume, is allowed five pages! In date the *Farbenlehre* is a century later (1810) than Newton's *Opticks*, but a reader might well suppose the reverse. The *Farbenlehre* is far more than a century out of date; its style is at times tiresomely laboured. 'We compare the Newton colour-theory with an old castle, which its builder', etc., for two octavo pages of small print, extending the 'Burg' simile.[1]

[1] *Zur Farbenlehre* p. 7.

Taken as scientific discussion its manner, even in its own time, is that of a past age. Nor was this an isolated instance in Goethe. In his *Annalen*, referring to the refraction of light, he wrote 'For here was the citadel of that bewitching princess who in an array of seven colours had befooled the whole world; here lay the grim sophistic dragon threatening everyone who presumed to try his fortune with these illusions'.[1]

Newton's *Opticks*, on the other hand, whether in its English or its Latin, has the speed, terseness, and precision of a first-rate modern thesis, over and above its discoveries of lasting value to the world. 'In his work on optics Newton introduced a fundamentally new point of view which men even as late and intelligent as Goethe completely failed to understand.'[2] The title-page of the *Opticks* does not mention the author's name; 'the only explicit mark of authorship being the letters "I. N." at end of the advertisement'.[3] Modesty? 'I do not know what I may appear to the world; but to myself I seem to have been only like a boy playing on the sea-shore, and diverting myself, in now and then finding a smoother pebble or a prettier shell than ordinary, while the great ocean of truth lay all undiscovered before me.'[4]

Were it not for Goethe's poetry, surely it is true to say we should no longer trouble about his science. Such as it was, it is as science not important. Its importance lies in

[1] And *Annalen*, p. 349.

[2] E. N. da C. Andrade, *Proc. Physical Soc.* (1943), vol. LV, p. 140.

[3] Ibid. p. 142.

[4] Newton, shortly before his death [Charles Singer, *History of Science*, p. 248].

the light it throws on Goethe the poet, and on his conception of Nature. It documents him a poet-pantheist. He thought about Nature over and over. He abounded in originality. His enthusiasm as an observer of Nature was great. But a new fact he met with was apt to send him on a flight of imagination into the unknown. Creative genius in literature, in science his genius longed to create. It could not always abide the waiting for further experiments and more knowledge. Science has to follow experiment where possible, even where the imagined seems extremely probable. Goethe, though devoted to science, had not at root the scientific temperament. He had not, for instance, along with the urge to discovery the sublime detachment of the scientific thinkers.

To compare an animal's skull with man's and point out the larger brain-case of man was already in the eighteenth century a standard piece of moralising—witness Oliver Goldsmith's *Natural History*. Goethe, conversing in similar strain (when he was over 80), contrasted with the animal's skull 'man's eminence'.[1] But to read into that remark a forecast of the theory of Evolution is mere obsession. As well suppose that Linnaeus, classing man and ape together as Primates, anticipated the *Descent of Man*. Goethe was intrigued by the fluctuations of detail noticeable in plant and animal, which the ardent systematists of his time tried not to notice. He thought of this fluctuation as Nature's failure to sustain fully the primordial pattern —Nature of course he humanised, much as the Greeks did the deities on Olympus. He had no suspicion of any connection between systemic classification of plants or

[1] *Gespräche m. Eckermann* (Oxenford trans.; Dent and Co.), p. 388.

of animals, and relationship by common ancestry. 'The ornithologists', he said, 'are probably delighted when they have brought any peculiar bird under some head; Nature however carries on her own free sport, without troubling herself with the classes marked out by limited men.'[1] 'The essential Form, with which Nature does but keep playing ever, so to say, and, playing thus, brings forth the manifold life.'[2] But Goethe never suggested that these errant forms led anywhere. In fact he said explicitly that they did not lead anywhere. He wrote there is 'in Nature an endless living, becoming and moving, and yet she gets no further'.[3] 'She seems to stake all on the individuals, and she makes nothing out of individuals.'[4] Goethe does not even ask whether these variants breed true. Of evolution there is in Goethe neither here nor elsewhere any hint. Witness also his biological comments on the Biblical account of the Creation.[5] In his figurative language he pictures Nature producing simply for *joie-de-vivre*. Lamarck in Paris was, however, at this same time arguing the progressive adaptation of form to purpose; but he had not reckoned with the non-inheritance of characters acquired by use. Darwinism of course came more than a generation later, and in Goethe, though he has been searched for it, there is no unequivocal hint of organic Evolution—perhaps in Keats, addressing the sky-children 'For on our heels a fresh perfection treads' (1825).

Preoccupation with heredity seems never to have been

[1] Ibid. p. 238.
[2] Goethe to Charlotte v. Stein, 10 July 1786. Duntzer, *Life*, p. 346.
[3] *Ges. Werke*, vol. XXX, p. 425.   [4] Ibid.
[5] *Gespräche m. Eckermann* (1828), p. 266.

part of Goethe. His marital relations suggest a like dis-regard.[1] Even the distinction which common parlance has long expressed as 'nature and nurture' had little place in his mind. He attributed variety and richness of growth on the luxuriant southern Alpine slope to fertile soil and smiling sun.[2] He seems to have accepted without question that 'nurture' was the formative influence. The *sans culottes* of the Revolution were to him a hopeless pro-blem as well as a repellent mob. Thomson's poem on 'Liberty' failed because 'there was no poetry in the theme "Liberty"'.[3] The cry *égalité* was hateful to him, and he grieved that France, his admired France, should be loud with it. Could he have known he would have longed for the sermon which twins 'identical and non-identical' preach to us to-day. The false slogan that men are born equal had no supporter in Goethe. Did it date from Jean Jacques, as some social platitude? Excusable perhaps in the welter of politics it is in biology a reprehensible un-truth. In biology it is so plainly untrue as to suggest intentional falsity. Biological equality among the members of a family, of a community, of a race, of a nation! Why, what does a stud-book mean? Has the stock-breeder no method and no aim? Have intelligence and physique no pedigree? And to-day, when knowledge can distinguish in the molecules of life those specifically fraught with heredity, shall we omit to study and use them? Heaven's great gift of *inequality*, allows the betterment of our kind and of the Earth. Now that we can see and know the

[1] *Gespräche m. Eckermann* (1824), p. 31.
[2] *Die Italienische Reise* (1816), vol. I.
[3] *Gespräche m. Eckermann*, p. 425.

very particles of life on whose reactions depend the differences among men, shall we neglect to employ them and to control them? Shall we still be content with chance and fortuity? Hybridisation and selection are laborious methods but reliable, and they require careful planning and carrying through. A factual difficulty is that to plan for the best we must know what the best is. A stock-breeder wanting more milk, can breed for more milk; but as an item in man to better the community the first points we think of are perhaps intelligence and probity. How shall we begin? Goethe is no help to us; he had no such ideas.

The category of 'naturalist' into which he falls is a different one. 'I have at times to resort to pantheism to satisfy my being.'[1] Pantheism always has had an Olympus. So with Goethe, for him his 'Nature' is an Olympian figure. He, a 'visual', visualised her; he saw her a creative Being, coeval with life, a mythical *type*, rather than any *character*, dramatic only as a Greek chorus may be dramatic. To personify 'types' was a poetic licence which had become a chronic ailment in European poetry. To personify 'types' became a temptation to Goethe as poet. There is a riot of it in the second part of *Faust*—queues of incidental figures uttering each some type 'point-of-view'; Nereids, lemurs, Dryads, Sirens, homunculus, Thales, Fear, Hope, Klugheit, etc., until such proper motion as the mythos itself has—and it has little enough—is clogged by this mob of 'personae'—and the myth itself is brought to a halt. Not that there are no fine passages; but they lie buried in a 'Walpurgis Night' of medley and irrelevance. An ordinary reader, such as myself, finds this

[1] *Gespräche m. Eckermann*, p. 425.

33

long procession of impersonated 'points-of-view'—some of them appropriate for delivery by Goethe himself rather than as allotted—tediously ineffective.

Some of them so lack differentiation as to be inter-changeable one with another, thus, *Mephisto* and *Phorkyas*, the double *Thersites*, *Boy Lenker* and *Euphorion*. Faust himself is throughout a 'type', not a 'character', no guilty living 'character' can be without remorse, as Faust is. One meets the opinion that in the *Faust* Goethe's nearest to characterisation is perhaps the Gretchen of Part I—aside from the prison scene. But even so he unloads on the young cottage girl the gorgeous '*Thule*' poem, out of keeping with her cottage simplicity.

Goethe regretted, to Eckermann, that not one of his songs lived, as for instance did those of his contemporary Burns, among the 'people'—'mine are favourites merely with pretty girls who play the piano'.[1]

In Part II he manufactures interludes in order to insert his own *obiter dicta* under recondite pseudonyms. The interludes do duty as cover for interruptions, from near and far, thus, feminine loveliness confronting its extreme antithesis, Cyclopean hideousness; or to put across an airy allusion, e.g. to Lord Byron—a Byron, save the mark, not longer flesh and blood. Or, perhaps more strange in-trusion still, *Homunculus*,[2] a pigmy false-birth in a flask, '*nur halb zur Welt gekommen*', a hoarseness in a bottle, lament-

---

[1] *Gespräche m. Eckermann* (1827), p. 202.

[2] The alchemical attempt to create man used to be set for the Doctorate-theses in the seventeenth century—e.g. at the Paris Faculty. *An in Vasis distillatoriis Homo possit creari?* M. Joan de Montigny, XXVII, Januar, Anno 1649. *Quaestiones Medicae*, Paris, 2⁰, 1752.

able and risible alike, but resurrected through a couple of scenes for a serious role. Humour was not among Goethe's gifts; for example, the successful contemporary satire 'Vision of Judgement', on Southey's somewhat pompous poem, remained to Goethe 'incomprehensible'.

The celestial setting of the Faust-world startles at first by the minute scale on which it is drawn. We remember, of course, that the sensible world has opened out vastly since Goethe wrote. Time has now been stretched to a dimension for 'relativity'. Space, in so far as we can take it by itself, has been expanded, and with it has enlarged our world. Our world as compared with that when Goethe wrote is not merely a thousand times larger, but, let us say, repeating Tennyson, a million million times larger. And, we have learned—and it saddens some—that of the vast world beyond the sun the more our knowledge enlarges it, the less human does it become. '*Comme nous sommes seuls pourtant sur notre Terre.*' I like to think it was part of Goethe's artistry to keep the Faust-world—*Der Herr*, the Archangels, the sun, and all such furniture of Heaven as he gives us—to puppet-show size, in order that it appear to us more human. He gives us, too, a strange sun which, instead of coursing through void, 'thunders along' as if in air! As for the Earth it remains the old Earth of Adam and Eve, now populated by their children's children, and especially by one family upon whom Mephisto intrudes, bringing Faust; Mephisto being depicted in the likeness of 'God's ape'. All this is supremely told, nor is it in Part I impaired by complication with numbers of those less significant allegorical figures to whom Goethe was unable to resist giving stage-room. In

35

Part II, however—he was an old man then—he yields to them in greater numbers; as said above, they clog the proper motion of his drama, they confront the spectator with, not *characters*, but speaking shadows.

For the man in the street the world in Goethe's time was measurably smaller than is the world to-day. Many had, of course, long discarded the idea on which they had been brought up that we good people are ourselves the central object for which this world is.

Mention being made at table of the efforts of certain inquirers into nature, who, to penetrate the organic world, would ascend through mineralogy, 'that', said Goethe, 'is a great mistake. In the mineralogical world the simplest, in the organic world the most complex, is the most excellent. We see, too, that these two worlds have quite different tendencies, and that a stepwise progress from one to the other is by no means to be found.'[1] This pertinent remark suggests that the question how far the furniture of Earth's surface, in sum, the non-living and the living together, is a single related series, was one not strange to Goethe's thought.

Goethe had more than most men the opportunity to 'edit' himself, and he took it. The *Gespräche*[2] mention his editing for reissue *Frankfort Literary Notices* written rather more than fifty years before. The stories of himself in his reminiscences are not all reliable.[3] *Wahrheit und Dichtung* was their fitly chosen title; and the *Annalen* show a similar trait. He was to himself, more than are most of us to ourselves, partly an imaginary character.

[1] *Gespräche m. Eckermann* (1831), p. 391.
[2] Ibid. p. 3.   [3] Cf. *Annalen* (1811), p. 399.

Goethe was—although he could not tell a lark from a sparrow [1]—from all evidence a devoted and even an impassioned lover of Nature. There was an impression abroad that as one man to another he was apt to be somewhat restrained and cold. However that may be, the perceptible world remained to him a thing of beauty, and the pageant it presented an unfailing delight. He seems sometimes, as we listen to him, to be looking over the shoulder of a creative Being, and to be entranced by watching her amid her work. In his earlier middle-age he was moved to set forth in a remarkable rhapsody [2] some thoughts on Nature's role in this world—a world which he there labels immortal. Among her creations he includes along with all the rest, ourselves.

We are in her and she is in us. Unasked and unwarned we are caught up by her into the whirl of her dance. She carries us along until we are tired and drop from her arms —she herself is tireless. Her present is eternity. Always is she a whole and yet never is she complete. All the time she constructs and she destroys all the time. Life is her fairest invention, and Death is her device for getting more life. She sows wants because she likes movement: The game she plays with all is a friendly game. The more one takes from her the better she is pleased. But she loves herself; her study is herself and her own pleasing. She invites us to share her own enjoyment of herself. She loves illusion. She shrouds man in mist, and she spurs him towards the light. Those who will not partake of her illusions she punishes as a tyrant would punish. Those

[1] *Gespräche m. Eckermann* (1827), p. 224.
[2] *Ges. Werke*, vol. XXX, p. 313.

who accept her illusions she presses to her heart. To love her is the only way to approach her. With her love-potions she can heal the woe of a life-time. She has brought me here; she will take me hence. I trust her. She will not hate her own handiwork.*

But that mood was not Goethe's always. Moods passed over him like brightness and shadow over a spring meadow, and he would revise his judgements and then praise both sides. Sibylline equivocality never entered him. The discrepancy between his reaction to nature in the 'Fragment' above quoted and in the following more widely known short poem *Das Göttliche* (1782) is surprising. It is not a difference between youth and age, still less between liberalism and toryism; he was a staunch conservative all his days, a 'friend of the established order', as he said. The theme of *Das Göttliche* is, despite its name, not 'the divine' in any ordinary sense. Its moving lines address that hegemony of mortal thought which we identify as human. Here are scraps of it:

> For Nature is callous,
> The sun shines above us
> On vileness and virtue,
> And on the wicked
> As on the worthiest
> Beam moon and stars.
>
> Wind and water-course,
> Thunder and hailstorm,
> Sweeping on headlong
> In one confusion,
> Hurl them together.
>
> *   *   *   *

Ruled by eternal
Vast awe-compelling
Ties that constrain us,
We must still onward
Tracing our being's
Ultimate ambit.

Man is the only
Crowner of goodness,
And he the only
Scourger of evil;
He heals and rescues
Those are in error,
Those that are straying,
Fain to be bidden,
Gathers he to him.

(Alas, my clumsy Englishing ruins it!)

It goes on to tell us that we human beings do well to model the Gods on man, and to make them prototypes of ourselves. In truth it equates man with the concept of God. I had wished the above two samples to illustrate this strange twin-ship of ideals, Man and Nature. There is, I think, no parallel to it in other mature poets of the time, but the Greeks had it.

We are animals. Not rarely we forget that—we become so taken up by the human part of affairs. But with Goethe we can fancy the fact was never long out of mind. He looked round upon the world about him, mindful that he was a sample of Earth's supreme and most knowledge-able product. His self-complaisance argues that he was well content with that status in his sum of things. From it he could bridge at a glance the gap between discordant

Nature and Man. An animal which chronicled its own doings, kept a census of its own numbers, and 'looking before and after' envisaged death, in short, partly understood itself and its world. Above all, an animal which was developing, sporadically and unevenly it is true, a new 'value' confronting 'self-preservation', the ancient safety-principle conserving life, life's sheet-anchor in so many dangers hitherto. A new 'value' was arising which could on occasion substitute itself for the long current 'self-preservation' urge; a novel impulse, sacrifice of self's life for the sake of another's life. In a word *altruism* had come to be. And in the human animal it was of an order other than elsewhere, much as the human animal itself is of that other order. How was it with Goethe in respect of this supremest human gift?

There attached to Goethe in the view of some who knew him the reproach of selfishness, and was well known enough for a biographer to mention it, in order to refute it.[1]

Goethe's was not an altruism which gives life itself to defend friends and country. His was, that is, not the patriot's or the soldier's altruism, and this lack cost him some friendships. But he had his large-heartedness none the less. When his mother became a widow, relatives wished to put her under legal guardianship because she was spending too much, but Goethe intervened and said, 'No, she has the right to spend all her property away if she desires, she has suffered long, and with noble patience, under a weary lot'.[2]

[1] Duntzer, *Life*, art. 47, p. 745.
[2] Ibid. bk. vii, ch. iii, p. 538.

In Rabelais,[1] almost hidden among grossness and quizzical irreverences, we suddenly find: 'Late at night, before going to their beds they went out to look at the sky....They prayed God, the Creator, adoring Him, ratifying their faith, glorifying His goodness that knows no bounds;...Then they went to their repose.' This calm break in the rarely broken riot of laughter of Rabelais, can apply in essence and conversely to Goethe. A whiff of *Sturm und Drang* disturbing the self-complaisance of Grand Ducal Weimar.

In Goethe's pantheism there is at times a streak of heartlessness. Wordsworth was in his time a pantheist, but never heartless, perhaps because less open-eyed about Nature. Wordsworth was, as are so many, reticent about the hateful in Nature. He acclaimed her even as one of Man's *moral* teachers; amid passing things, 'The soul of Beauty and enduring Life'.[2] But, Nature is Janus-like, a gateway to two worlds, and unlike Janus-gate with both sides always open.

In that earlier rhapsody, on 'Nature', Goethe sang, 'Death is her trick for getting more life'; the game she plays with all is a friendly game. The game is 'Zest-for-life'. Zest-for-life! The most ruthless of all life's destroying angels. Goethe's thought runs wide of much of our sentiment of to-day. A superhuman Being launching new lives for the pure delight of doing so—lives whose purpose is still to seek—may be true up to a point; but it can become a grim story. This Being, if 'one of the sky-children', can be perverse. He calls her 'Nature'. For the

[1] John Cowper Powys, *Rabelais.* (London: John Lane, 1948.)
[2] *Prelude*, vol. VII, p. 736.

human observer, Nature in many of her aspects is beautiful. Sea, land and sky can have overwhelming beauty; so, too, the world of living things with their forms and colours and movements. To perceive Nature's beauty is perhaps among the loveliest of all human privileges; a quality of peculiar charm because no two of us can enjoy it wholly similarly. All beauty is relative, and for each of us this world is the only world we know. But, beside the lovable, our world has in it also the hateful, the abominable, vice, ugliness, blood-lust and cruelty. To take subhuman nature: the ferret gnawing the earthed rabbit, the wolf-pack in pursuit, the leopard at the throat of the deer, and then, the fever-fit of ague, the growing cancer, the whole scheme of destruction from the killer-whale to the ichneumon fly. This and much more, not to speak further of human cruelty and hate. Goethe, the amateur of Nature, must have known these facts, he must have perceived that, in the scheme of Nature, to attack and prey upon others is for the majority of animals, even likest ourselves, their charter of existence. Goethe had read Aristotle, the great master, and his summing up: that Nature taken all in all is more unlovely than lovely. He would know that the gifted pre-Christian singer of Nature had sung 'Nature is no product of the Gods, nor is it any gift to man: the spots which blacken Nature's face are too hateful'.[1] That leagues on leagues of verdant tropic forest are death-traps to civilising man and to harmless beasts of the field is wholly in keeping with one side of Nature. That Nature is bad as well as good is not a thing to be forgotten; civilisation tries to impose new

[1] *De Rerum Nat.* v, 198.

42

rules on 'zeal-to-live'. Zeal-to-live is an innate exhortation, but it is also a tyrant's inexorable decree; with progress of knowledge that fact emerges only the more clearly. Nature sows and reaps vast harvests of pain and fear all over our planet, on earth, in sky and sea. Yet turning to Goethe and his rhapsody it would seem unknown to him. It casts no cloud for him, although his great younger contemporary was protesting 'it spoils the singing of the nightingale'.[1] It forms part of that acknowledged problem, the co-existence with us of pain and suffering, physical and mental, animal and human. Yet Goethe says of Nature 'mit allen treibt sie ein freundliches Spiel'.[2] Ein freundliches Spiel! The words seem laden with irony. 'She is too lofty to heed their suffering though they are her own offspring! To do so would not beseem her!' Strange pride! Goethe's view of Nature remains in this respect an enigma, and it is another view than that of to-day.

Goethe writes in his quasi-autobiography:[3] 'In the course of this biography we have circumstantially exhibited the child, the boy, the youth, seeking by different ways to approach to the Super-sensible, first looking with strong inclination to a religion of nature.'

Spinoza, who had Goethe's ear in philosophy, has been called pantheist and also God-intoxicated. Goethe at one period in his worship of Nature seemed Nature-intoxicated. He may in later years have seemed so to himself. Confronted in old age with his earlier paean to the

---

[1] Keats, to Reynolds, l. 85.
[2] *Ges. Werke*, vol. xxx, pp. 9, 426: *Natur*.
[3] *Gespräche m. Eckermann* (1832), p. 422.

43

Nature-goddess he remarked that it did not represent his 'final' view. He wrote that it failed to reckon with the great driving force in nature—'polarity'. 'Polarity' as a specific universal force in nature has failed to establish itself either in science or philosophy; it remained a suggestion, unfortified by argument of any kind. But it was characteristic of its author. Of Goethe as a young man his friend Johann Kerstner, who met him first in 1772, and drew his portrait, has left a brief but vivid outline of impressions which Goethe made on him at that time. 'He possesses an extraordinarily active imagination, so that he speaks mostly in images and similes. He used to say that he never could express himself exactly. He holds Rousseau high, but is no blind adorer. He is not what is called orthodox. He opens his mind on great subjects only to a few. He would fain not disturb others in their tranquil beliefs. He hates scepticism. He goes neither to church nor to the Lord's Supper, and rarely prays; "I am", he says, "not hypocrite enough". He reveres the Christian religion, but not in the form in which our theologians present it. He *trusts* there is a future life. He has every kind of knowledge but that which earns bread.' [1]

The almost inexpressible immensities which to-day challenge our faith in human importance, were less oppressively within the knowledge of Goethe and his time than they are of ours. Changes more important than he knew had been and were going forward in science. As regards science, he himself, although he did not suspect it, lay becalmed so to say, in a small quasi-scientific back-

Duntzer, *Life*, p. 146.

water of no great account. But Boyle had been starting modern chemistry, and Lavoisier, Cavendish and Dalton had established it. Lagrange and Laplace had annexed the nearer heavens for Newton. Joule would soon be gauging quantities of 'heat' against quantities of 'work'. The quantitative liaison between the several material forces was still partly to seek, nor were certain of the forces themselves clearly recognised. Goethe dealt with some of these by notions *ad hoc* out of his own fancy, for example, *Polarität* which resisted *Entzweiung*; *morphogenesis*, yearning after some divine pattern. The very names suffice to indicate how remote his thinking was from the positions science had already reached. As Helmholtz said, 'Goethe's statements do not fall even within the province of natural science'.

Further, Goethe's pronouncements were shot through with anthropomorphism, which, unless used purely as metaphor, *is* the occult. In Goethe it *is* the occult. Polarität, morphogenesis, and so on, have 'yearnings', 'likes', 'will', etc. We are back in the mediaevum and early renaissance. Goethe advised Schiller to insert the astrologer in *Wallenstein*—he was right, but it was characteristic of him, as well as right.

The world must all of it if, as we suppose, it is a running concern in itself, be truly to scale; did we but understand enough, it would be obvious to us. But only a part of it is measurable by us—the physico-chemical. If measurement is the beginning of human knowledge, what are we to do? Thought is what we particularly want to measure. It is lack of precise measurement there which keeps our knowledge from beginning.

45

Rough estimates have to suffice. The range of size of lifeless things is far wider than that of living things. The range of size of those living things invested with psyche is much narrower than of lives in general. The range of size of lives exercising rational choice and reason is more restricted still. To imagine an intelligent being of microscopic or submicroscopic size seems to first thought, although man starts by being microscopic, almost a contradiction in terms. And it *may* be so. The doubt shows how slight our knowledge of the matter is. There is an English essayist who used to insist at some length on the unimportance of size. If the universe is to scale, that itself refutes him. The living world is, like it or not, a world of natural inequality. Inequality of size is part of that inequality. Upon that field of inequality is sown broadcast the seed of zest-for-life, and not merely broadcast and at certain seasons but in superabundance and always. The prize offered? Individual existence. The competition? Internecine. How successful man is biologically is perhaps best answered by remembering that he substantially is the only one of Earth's creatures permitted to die of old age— and not altogether rarely does so. The ocean is more populated than the land, and the naturalist tells us that it is unlikely that any individual life in all the oceans ever dies a natural death. I do not know whether Goethe, with his gift of picturing Nature's problems, guessed—he could not do more than guess—the extent of this competition. He certainly, walking this green planet, and adoring his imagined goddess, observed this universal struggle and endorsed it to a degree that Christianity never has observed it or endorsed it. To him the crown of Nature's god-head was her insatiable 'zest-for-life'. He said, 'Let people serve Him who gives to the beast his fodder, and to man meat and drink as much as he can enjoy. But I worship Him who has infused into the world such a power of production, that, when only the millionth part of it comes out into life, the world swarms with creatures to such a degree that war, pestilence, fire and water cannot prevail against them. That is *my* God![1] We recognise in this something of the rhapsody on *Natur* of fifty years before. It was with a virile joy that he beheld Earth's fertility and he applauded it as part of the divine activity which he, denizen of the same planet, was privileged to witness, and, in the van of Earth's creation, in part to understand.[2] His attitude had in it something of that of to-day's poet toward the bullfinch in the winter wood,

> His miracle's strange rightness.[3]

—'his miracle's strange rightness'. Finely and wisely said.

Goethe was a fortunate soul; he always had an enthusiasm ready, except for the French Revolution[4] and for poems 'snatched from the air'.[5] A change which did seem to come to him towards his end was greater concern about how other people felt regarding him. He issued a poem entitled '*Kein Wesen kann zu nichts zerfallen*'[6], perhaps to exhibit his good will to all men. We find him

---

[1] *Gespräche m. Eckermann* (1831), p. 389.

[2] Duntzer, *Life*, bk. III, cap. 5, p. 200.

[3] Claude Colleer Abbott, *The Bullfinch*, in *The Sand Castle and other Poems* (London, 1946).

[4] *Hermann u. Dorothea.*   [5] *Gespräche m. Eckermann*, p. 8.

[6] *Ges. Gedichte*, p. 290.

protesting (to Eckermann) against the rumour that he was 'without Christianity'. *A propos* of a New Testament, which Eckermann purchased and was showing him, he, among other comments, remarked[1]

there is in them [the four Gospels] the reflection of a greatness which emanated from the person of Jesus and which was of as divine a kind as ever was seen upon earth. . . . I bow before Him as the divine manifestation of the highest principle of morality. If I am asked whether it is in my nature to revere the sun also, I again say "certainly!" for he likewise is a manifestation of the highest Being, and indeed the most powerful that we children of earth are allowed to behold. I adore in him the light and the productive power of God; by which we all live, move and have our being—we and all the plants and animals with us.

In the *Annalen* dating when he was 62, Goethe declares 'with me, a pure, deep, innate and constantly followed conception has been the view that God is inseparably within Nature and Nature inseparably within God,[2] and this idea has been at the basis of my whole existence'. On another of those rare occasions on which he spoke of religion as regards himself, he is reported as saying that the Deity as such was to him inscrutable, but that one aspect of the Deity was not so, Nature. When he was 73 he received a letter from his friend of long standing, Countess v. Bernstorff (*née* Stolberg), after his dangerous illness. It concluded by urging him to turn his heart to eternal things and to make good, ere not too late, the harm his

[1] *Gespräche m. Eckermann*, p. 422.
[2] Ibid. (1824), p. 35.

works had done to the souls of others. Goethe answered her thus:[1] 'All my life I have meant honestly towards myself and others, and in all my earthly action have looked to the Highest. You and yours have done the same. We will then continue to labour while it is day; a sun will shine for others also. And so let us remain untroubled about the future. In the Kingdom of our Father are many provinces.'

Goethe was granted an old age which suffered little from the usual disabilities. His daily interests, instead of thinning away, remained many. New plays, public events, plans for a harbour to be, a fresh medal for his cabinet, another Waverley novel, social hospitality, besides the continued publication of his collected works, and instructions to Eckermann with superintendance of editing, etc., each day was a full one. We may picture him as loving the new, while still treasuring the past; to-day television, the wireless, etc. would we may think have fascinated him.

He still holds by 'God in Nature and Nature in God'[2] as he did more than thirty years earlier, but now he has something to add.

Let anybody only try with human will and human power, to produce something that may be compared with the creations that bear the names Mozart, Raphael, Shakespeare. God did not retire to rest after the well-known six days of Creation, but is evidently as active as on the first. It would have been for Him a poor occupation to compose this heavy world out of simple elements

[1] Duntzer, *Life*, p. 644.
[2] *Gespräche m. Eckermann*, p. 405 (1811) and *Natur* (1782).

and to keep it rolling in the sunbeams from year to year if He had not had the plan of founding a nursery for a world of spirits upon this material basis.

When Goethe said this he knew that the great change could not be far distant from him. What that change meant he was too wise to think he knew. But as a poet he approached it with we may suppose a poet's imagination. The valiant *Stirb' und Werde* was no empty boast with him. In the shadow of what was drawing nearer, among thoughts which lent him courage, one which seems to have grown to conviction was that the human spirit had Earth for its nursery. Our planet had cradled Mozart, Raphael, Shakespeare; these figures rose before him as samples of the human 'motif' at its finest. They were leaders by reason of what they had in their turn created. In them the human spirit had achieved what the divine spirit had desired of it. They had been leaders of 'man'; nor were they all there were. Can we doubt he was in thought adding to their names his own? With little of his life remaining it gave him content to add his own name to those others who formed the choicest stock and essence of the human spirit. That was a recompense for death's blow. Death he felt, while it cut the breath short, set in a sense its seal on such quasi-immortality as a man could earn. Fame was but oblivion postponed, but it was most assured attained perhaps at the grave's edge. He felt even at the grave's edge that he for one would remain long an unforgotten part of Earth's endeavour. With that as consolation he could, whatever life's end meant, face it without flinching; 'es kann die Spur von meiner Erdetagen nicht in Äonen untergehen!'

NOTES

Page 6

Magnus, Rudolf. *Goethe als Naturforscher*. Vorlesungen gehalten im Sommer-Semester 1906 an der Universität Heidelberg, von Rudolph Magnus....Mit Abbildungen im Text und auf 8 Tafeln. Leipzig, J. A. Barth, 1906.

Page 12

'Faithful observers of Nature, even if in other things they think very differently, nevertheless agree together that all which appears, everything that we meet as a phenomenon, must either mean an original division which is capable of union, or an original unit which can be split and in that manner exhibit itself. To sever the conjoined, to unite the severed, that is the life of Nature; that is the eternal drawing together and relaxing, the eternal syncrisis and diacrisis, the taking in—and the pouring out of breath of the world in which we live, and move and are'.[1]

Page 13

Tyndall, John. 'The Theory of Colour.' *The Popular Science Monthly*, 1880.

Schuster, Arthur. *Goethe's Farbenlehre*. Publications of the English Goethe Society, No. v, pp. 141–51, 1890.

Ostwald, Wilhelm. *Goethe, Schopenhauer u. d. Farbenlehre*. Leipzig, 1931.

Page 14

Among instances with which Newton had dealt was the colour of the rainbow. Goethe, in even the last months of his life, entered upon preparation of a new explanation.

[1] *Zur Farbenlehre*, vol. xxviii, Abt. 4, par. 739, p. 187.

Eckermann, writing in 1829, says 'Goethe could not readily bear contradiction with respect to his Theory of Colours. His feeling for the Theory of Colours was like that of a mother who loves an excellent child all the more the less it is esteemed by others'.[1] Eckermann had to admit that in certain matters he could not confirm the Theory.[2] Goethe was much disappointed at that. The year before his death, speaking of the lack of converts to the theory, he explained that it was one difficult to communicate to others.[3]

He writes suggesting that Newton has concealed ('verbirgt uns') part of an experiment! He can have studied the dead Newton to little purpose; such trivial trickery was no part of Newton. We are left wondering how Goethe could suppose it. A different ethic? Yes, but—Goethe had his defects! Heine called him 'the old egoist'. Others have muttered 'the old flunkey!' 'Her Imperial Highness the Grand Duchess was graciously pleased to allow me to write some poetical lines in her homage in her magnificent album.'[4] His own lips have given samples of unworthy prejudice. But Napoleon addressed him, 'Vous êtes un homme!' Taken in the round, he remains Goethe.

Goethe seems to have grown annoyed by the way his Colour controversy was going. He began to feel it injured his scientific repute. In asserting himself he was impulsive enough to traduce the character of the dead Newton. He published a sketch of Newton's *Persönlichkeit*[5] 'er sei als Mensch, als Beobachter, in einen Irrthum gefallen, und habe als Mann von Charakter, als Sectenhaupt seine Beharrlichkeit eben dadurch am kräftigsten bethätigt, dass er diesen Irrthum, trotz allen

---

[1] *Gespräche mit Goethe*, p. 502.
[2] Ibid. p. 301.   [3] Ibid. p. 418.
[4] *Annalen* (1821).
[5] *Ges. Werke*, vol. XXIX, p. 217.

äussern und inneren Warnungen, bis an sein Ende fest behauptet, ja immer mehr gearbeitet und sich bemüht, ihn
auszubreiten, ihn zu befestigen und gegen alle Angriffe zu
schutzen.' 'Erst findet er seine Theorie plausibel, dann
überzeugt er sich mit Uebereilung, ehe ihn deutlich wird,
welcher mühseligen Kunstgriffe es bedürfen werde, die
Anwendung seines hypothetischen Aperçu's durch die Erfahrung durchzuführen. Aber schon hat er sie öffentlich
ausgesprochen, und nun verfehlt er nicht, alle Gewandtheit
seines Geistes aufzubieten, um seine These durchzusetzen;
wobei er mit unglaublicher Kühnheit das ganz Absurde als ein
ausgemachtes Wahre der Welt in's Angesicht behauptet.'[1]

In other words Goethe asserts that Newton, knowing
himself wrong, brazens it out; Goethe has not the slightest
ground for saying this. Further he goes on, Newton is quite
right, as leader of a controversy, to brazen it out—a grotesque
misreading of the situation and of Newton's character. Again,
writing of Newton's description of an experiment, he comments: 'Allein er übersieht oder verbirgt uns.'[2] In 1810 Goethe
vented his spleen against Newton's memory in some verses
entitled *Katzenpasteten* (cat-pies).[3]

Again, 'Nur nach dem kümmerlichen Anlass der Com-
pendien in welchen sich die Newtonsche Lehre, die doch
anfangs wenigstens ein Abracadabra war, zu unzusammen-
hängenden Trivialitäten verschlechtert.'[4]

Page 38

This remarkable composition known as '*Die Natur*'[5]
appeared unsigned in the twenty-second number of the

[1] *Ges. Werke*, vol. XXIX, p. 174.
[2] Ibid. vol. XXVIII, p. 333, par. 1222.
[3] Duntzer, *Life*, p. 598. *Ges. Gedichte*, vol. II, p. 90.
[4] *Ges. Werke*, vol. XXX, p. 12: *Nachträge zur Farbenlehre*.
[5] *Ges. Werke*, vol. XXX, p. 313.

Journal von Tiefurt. It was originally headed '*Fragment*'. Its original MS. is in the handwriting of Goethe's secretary and bears corrections in Goethe's own hand. The Jubilee edition of Goethe does not doubt Goethe's authorship of it. It dates at the end of 1780 or early 1781.

In 1829 it came into Goethe's life again, with things from the Duchess Amalia. Goethe then said it did not represent his 'final thinking' on its subject; his final thinking he said included the two great driving-wheels of Nature which were 'polarity' and 'enhancement' (*Steigerung*). We have already dealt with what he meant by these. The term 'polarity' as used in this sense by Goethe has no reference to light. 'Steigerung' stands for an enhancement of colour. Goethe was very careful to re-edit his early utterances—not merely the quite youthful. In the recently published, finely produced catalogue of the *Goetheana* at Yale University[1] I notice that their copy of *Die Natur* is classed among the *apocrypha*. But an original in the handwriting of Goethe's secretary, with emendations in Goethe's own hand! As a composition it is impressive and characteristic. His ultimate discussion[2] of it amounts to his admission of it as his own.

It is admitted without question to the 30-volume issue of Goethe by Cotta, Stuttgart, 1858.

[1] Most kindly furnished to me by Dr John F. Fulton of Yale.
[2] *Ges. Werke*, vol. XXX, p. 425.

Milton Keynes UK
Ingram Content Group UK Ltd.
UKHW032320161024
449665UK00001B/17